SLITHERING SNAKES

Written by Paul Stevenson

CONTENTS

Deadly Snakes	4
Fangs	6
Venom	8
Deadliest Reputation	10
King Cobra	12
Black Mamba	14
Deadly Squeeze	16
Swallow Hard	18
Warning!	20
Survive a Snake Bite	22
Snake Bite Cures	24
Snake Mothers	26
SOS - Save Our Snakes	28
Snakes as Pets	30
Glossary	31
Index	32

First published in 2024 by
Hungry Tomato Ltd
F15, Old Bakery Studios,
Blewetts Wharf, Malpas Road,
Truro, Cornwall,
TR1 1QH, UK.

Copyright © 2024 Hungry Tomato Ltd

No part of this publication may be reproduced, stored in a retrieval system, or transmitted in any form or by any means, electronic, mechanical, photocopying, recording, or otherwise, without prior written permission of the copyright owner.

A CIP catalogue record for this book is available from the British Library.

ISBN 9781916598775
Printed in China

Discover more at
www.hungrytomato.com

Neither the publisher nor the author shall be liable for any bodily harm or damage to property whatsoever that may be caused or sustained as a result of conducting any of the activities featured in this book.

DISCLAIMER:
The people who hunt and handle snakes and their venom are experienced professionals. Under no circumstances should you try engaging with a snake yourself, as it is extremely dangerous!

All words in **BOLD** can be found in the glossary.

DEADLY SNAKES

Every year, around 100,000 people die from snake bites! Most **victims** live in the countryside of Africa, India and Southeast Asia.

People die because they can't get to a hospital for treatment. The **venom** in the snake bite attacks their body before they can get help.

Indian cobra

The Cape cobra is one of Africa's most dangerous snakes. Without treatment, the venom in a single bite can kill a person in just a few hours.

Cape cobra

FANGS

Snakes bite their prey with two long, pointed teeth called fangs.

Venom is squirted through the fangs.

The snake's venom goes straight into its prey's bloodstream.

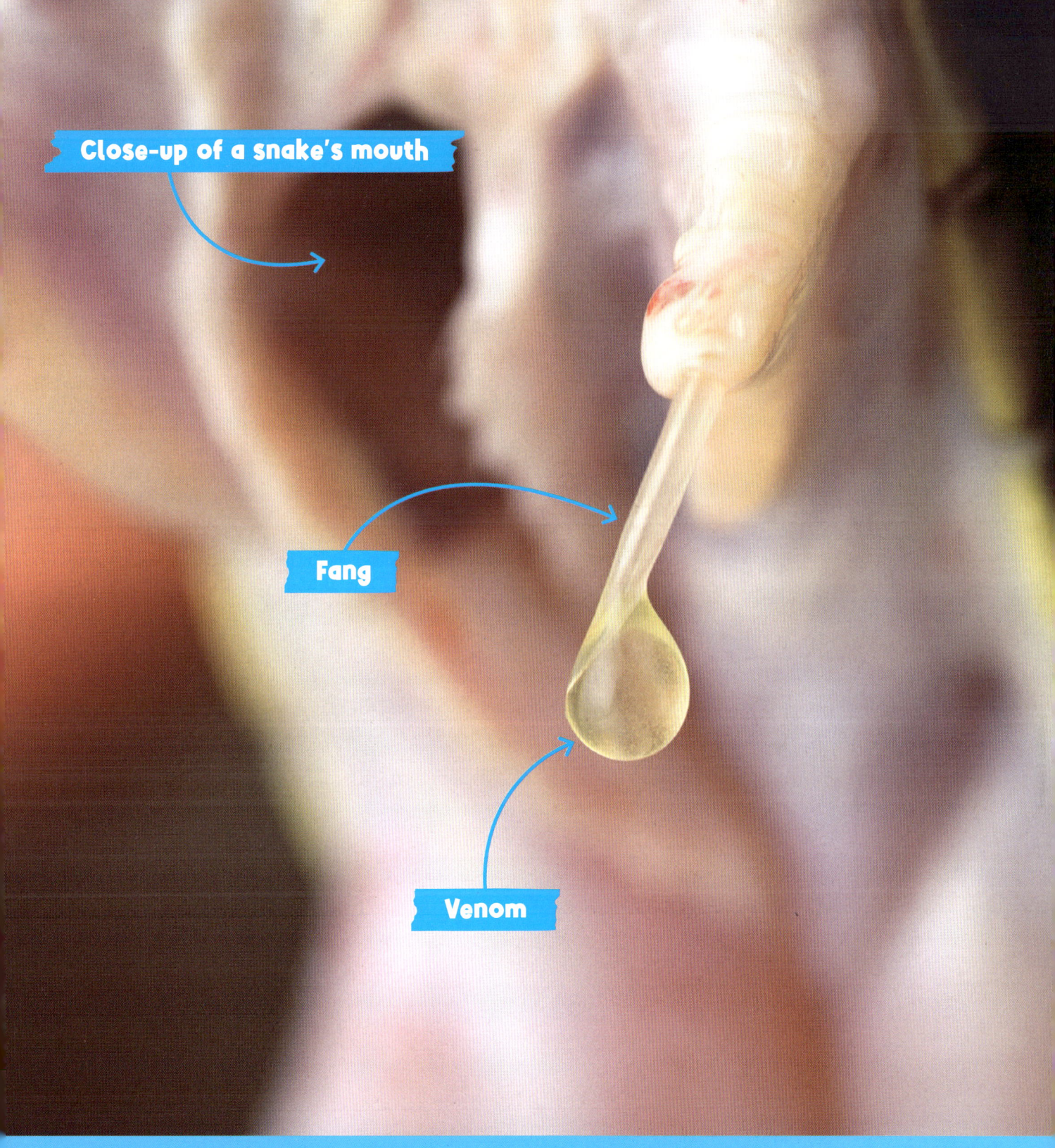

Close-up of a snake's mouth

Fang

Venom

CAPE COBRA VENOM IS STRONG ENOUGH TO KILL 9 HUMANS WITH A SINGLE BITE!

VENOM

There are two main types of snake venom.

Spitting cobra

Spitting cobras spit their venom into their prey's eyes. They can hit their target from 3 metres away!

NEUROTOXIN

How it works: it stops a victim's nerves working so they can't carry messages to their muscles.

What happens: the victim can't breathe. They will die within hours if they don't get treatment.

Some snakes with this venom: kraits, cobras, wagler's pit vipers, and mambas.

Wagler's pit viper

HAEMOTOXIN:

How it works: it stops blood from **clotting**.

What happens: the victim might bleed to death on the inside if they don't get treatment. Those who do survive may develop **gangrene** and possibly lose limbs!

Some snakes with this venom: pit vipers and rattlesnakes.

DEADLIEST REPUTATION

Snakes are normally scared of people. They only bite if they are disturbed and feel they are in danger, but they can be very deadly!

The saw-scaled viper may be the deadliest of all snakes, since scientists believe it's responsible for more human deaths than all other snake **species** put together.

It doesn't kill every victim, but the snake is aggressive; it bites early and often!

Saw-scaled viper

The banded krait may be a little more chilled out, but its highly toxic venom is a killer!

You can recognise a banded krait by its yellow and black stripes.

Banded krait snake

KING COBRA

The king cobra is the longest **venomous** snake in the world. It can grow to over 5 metres long.

This man is Khum Chaibuddee. In 2006, he set a world record for snake kissing. He kissed 19 king cobras!

This was either very brave or foolish considering the volume and strength of the king cobra's venom can kill an elephant in just a few hours!

THIS KING'S BITE IS NOT TO BE MESSED WITH!

When there is a threat, a king cobra will rise up to attack. Its head can be 1.5 metres above the ground.

BLACK MAMBA

Black mamba bites are always fatal, unless treated immediately!

Africa's longest venomous snake can grow up to 4.3 metres in length and can have a slithering speed of up to 12.5 miles per hour!

They normally stay hidden from people. However, hot weather in Durban, South Africa, in 2016 led to an infestation in residential areas. One man was bitten by a baby mamba but survived.

People can often find them in their gardens.
The rule is: keep your distance and don't provoke or try to catch it.

When people get bitten, it's normally when they are surprised by them!

Baby black mamba hatching

Black mambas are fast. They can slither faster than a person can run.

DEADLY SQUEEZE

Pythons and constrictors squeeze their prey to death.

They wrap their huge bodies tight around their victims.

Crocodile

16

Every time the victim breathes out, the snake tightens its grip.

The prey **suffocates** to death.

Rock python

SWALLOW HARD

Snakes do not chew their food. They swallow it whole!

First, the snake kills its prey using its venom or by squeezing its victim to death.

Next, the snake swallows its prey's head.

A snake's lower jaw is not attached to its upper jaw. This means it can open its mouth very wide.

Then, the snake slithers its own body over its meal until everything is inside.

Sometimes the snake's prey isn't completely dead.

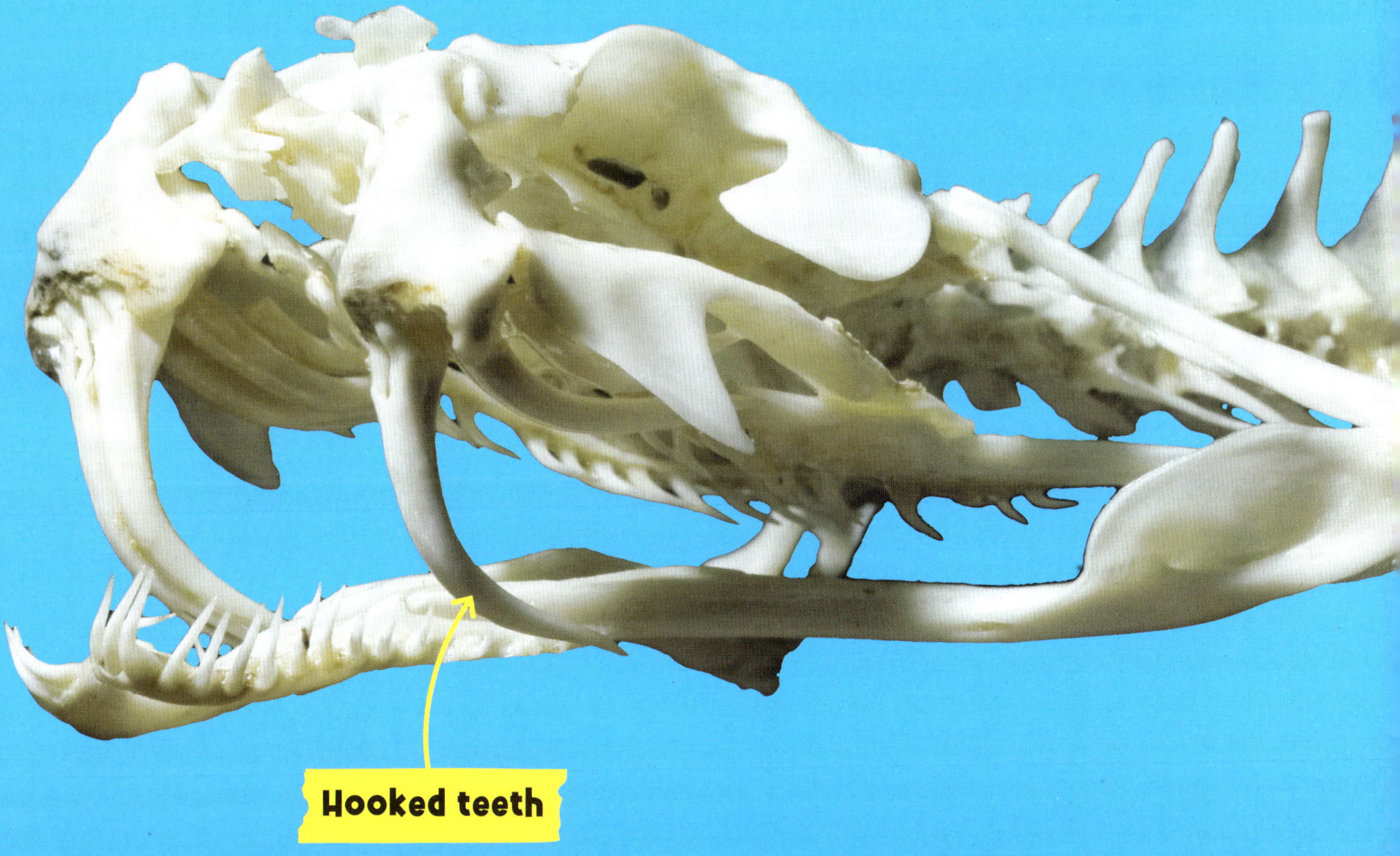

Hooked teeth

Snakes use their hooked teeth to hold the struggling prey as it is swallowed alive!

WARNING!

A snake's venom is its main way of catching prey and defending itself, so it doesn't want to waste it!

Most venomous snakes give a warning before they attack.

If you corner a rattlesnake, it will use the rattle on its tail to frighten you off.

The rattle is made of dead skin. It grows longer as the snake gets older.

Rattle

Rattlesnake

Venomous coral snakes have bright warning stripes.
But some non-venomous snakes have copied this.

A coral snake has red stripes touching yellow stripes.

A kingsnake is non-venomous. Its red stripes touch black stripes.

REMEMBER THE DIFFERENCE WITH THIS RHYME:

**Red touching yellow kills a fellow.
Red touching black, venom it lacks!**

SURVIVE A SNAKE BITE

WHAT TO DO IF YOU GET BITTEN:

- Move beyond the striking distance of the snake.

- Remain calm and still as this will help slow the spread of the venom.

- Remove jewellery or clothing before you start to swell.

- Try to position yourself so that the bite is below your heart.

- Clean the wound with soapy water and cover with a dry dressing.

- Remember what the snake looks like and call the emergency services or get to a hospital.

WHAT NOT TO DO IF YOU GET BITTEN:

- Apply ice or a tourniquet.

- Drink alcohol or caffeine.

- Try to capture the snake!

SNAKE BITE CURES

Snake bites are treated with medicines called antivenoms.

Antivenoms are actually made from snake venom!

Snake venom is valuable stuff. A gallon of venom can be worth millions of pounds!

Indian cobra snake

The man in this photo is trying to catch an Indian cobra. He makes a living catching snakes. He catches them so their venom can be collected.

Collecting snake venom is known as "milking the snake".

Venom

The venom is then injected into a large animal, such as a horse. Horses and sheep have strong **immune systems** that produce powerful **antibodies** when injected with small amounts of snake venom.

The horse's blood is then used to make the antivenom.

SNAKE MOTHERS

Snakes may be deadly killers, but many are caring mothers.

Eggs

A female python wraps her body around her eggs.

She shivers to make heat. This keeps the eggs warm until they **hatch**.

Green tree python

Baby snake

SOS – SAVE OUR SNAKES

Many species of snake are endangered. They need help!

People often kill venomous snakes because they are afraid of them.

Snakes also lose their **habitat** when people clear land to build homes or grow crops.

Zoo visitors holding a python

There is some good news. Zoos give snakes a safe place to live. They also teach visitors about the snakes.

Many zoos are **breeding** endangered snakes. Sometimes zoo-bred snakes can go back to the wild.

The Australian woma python is endangered.

Woma python

Zoo-bred snakes are often released back into the wild, in safe nature reserves.

Aruba rattlesnake

Many zoos and aquariums take part in breeding snakes to save endangered species, such as Aruba rattlesnakes.

SNAKES AS PETS

Would you like a pet snake? Think it through carefully...

- Many snakes can live for over 20 years. Do you want to live with your snake for that long? That's a long-term roommate!

- Keeping a python might seem like fun, but it could grow to over 3 metres long! Do you have enough room?

- Snakes are wild animals who belong in natural habitats. Wild animals in **captivity** experience some degree of distress at the loss of freedom (unless they were born in captivity).

Corn snakes make good pets. They grow up to 1.5 metres long.

GLOSSARY

antibodies - special proteins in an animal's blood that fight infections.

breeding - putting male and female animals together so they mate and have young.

captivity - living in a cage or enclosure, such as in a zoo, on a farm or as a pet.

clotting - when liquid blood becomes jelly-like and doesn't flow any more.

defend - to protect.

endangered - at risk of dying out so there are no more of that animal species left.

fang - a long, sharp tooth. Snake fangs are hollow so venom can be pumped through them.

fatal - something that is extremely dangerous and can lead to death.

gangrene - a very bad infection. Sometimes a body part has to be removed if it has gangrene.

habitat - the place where an animal lives. Snakes live in many different habitats, from forests to deserts.

hatch - to break out of an egg,

immune system - the body system that fights off illness.

prey - an animal that is hunted by another animal as food.

species - a group of animals that look similar and can breed with each other.

suffocate - when an animal or person dies because they can't breathe.

venom - a poisonous substance injected into prey by biting, stinging or spitting.

venomous - an animal that produces and uses venom to kill prey and defend itself.

victim - a person or animal who is hurt or killed.

INDEX

A
antivenom 24-25
Aruba rattlesnake 29

B
baby snakes 14, 26-27
bite 4-5, 6-7, 10, 12, 14, 22-23, 24-25
black mamba 14-15
banded krait 11
breeding snakes 28-29

C
Cape cobra 5, 7
Chaibuddee, Khum 12
coral snake 21
corn snake 18, 30

E
eggs 26-27

F
fangs 6-7, 31
food 18-19

H
haemotoxin venom 9
hog-nosed pit viper 6
horse 25

I
Indian cobra 4, 24

K
king cobra 12-13
kingsnake 21

M
medicine 24-25
milking 25

N
neurotoxin venom 8

P
pythons 16-17, 27, 28-29, 30

R
rattlesnake 9, 20, 29
rock python 16-17

S
saw-scale viper 10
spitting cobra 8

T
teeth 6, 19

V
venom 4-5, 6-7, 8-9, 10-11, 12, 18, 20-21, 22, 24-25, 31

W
Wagler's pit viper 8
woma python 29

Z
zoo 28-29

Picture credits:

(t=top; b=bottom; c=centre; l=left; r=right):
Shutterstock: 2-3, 9, 18, 19t, 19c; Kurit Afshen 1, 22-23, 30b; NickEvansKZN 14b, 15; Potray 8br; Roberto 33 12-13; Ryan M. Bolton 20; Stu Porter 8t. age fotostock / SuperStock: 10. Claus Meyer/ Minden Pictures/ FLPA: 7. FLPA/FLPA: 16-17. How Hwee Young/ epa/ Corbis: 11. Jeff Greenberg/ Alamy: 28. Jeffrey L. Rotman/CORBIS: 25. Michael & Patricia Fogden/ Minden Pictures/ FLPA: 6, 21tl, 21tr, 29t. NHPA/Daniel Heuclin: 4, 24. Photolibrary Group: 26-27, 31b. Rungroj Yongrit/ epa/ Corbis: 12. Tony Phelps/ naturepl.com: 29b.

Every effort has been made to trace the copyright holders, and we apologise in advance for any unintentional omissions. We would be pleased to insert the appropriate acknowledgements in any subsequent edition of this publication.